NICOLAUS COPERNICUS

GREAT ASTRONOMERS

ROBERT STAWELL BALL

GIFT *Certificate*

TO:

FROM:

DATE:

NICOLAUS COPERNICUS

GREAT ASTRONOMERS

BY

ROBERT STAWELL BALL

D.Sc, LL.D., F.R.S.

(July 1, 1840 – November 25, 1913)

LOWNDEAN PROFESSOR OF ASTRONOMY AND GEOMETRY
IN THE UNIVERSITY OF CAMBRIDGE

DIAMOND™ BOOKS

www.diamondbooks.ca

TORONTO, CANADA – 2016

DIAMOND BOOKS - CANADA

DIAMOND BOOKS

Toronto, ON, CANADA

http://www.diamondbooks.ca

BIBLIOGRAPHIC INFORMATION

'NICOLAUS COPERNICUS'
Is an extract or a chapter from the book
"GREAT ASTRONOMERS" by Robert Stawell Ball.

FIRST PUBLISHED IN: 1896.

PUBLISHED IN CANADA

Published in Canada by DIAMOND BOOKS - CANADA, an imprint of DIAMOND PUBLISHERS - http://www.diamondpublishers.com

FIRST EDITION: OCTOBER, 2016.

PAPERBACK EDITION : ISBN: 978-1-988357-30-0

PRINTED IN CANADA

PREFACE
TO THE ORIGINAL EDITION

It has been my object in these pages to present the life of each astronomer in such detail as to enable the reader to realise in some degree the man's character and surroundings; and I have endeavoured to indicate as clearly as circumstances would permit the main features of the discoveries by which he has become known.

There are many types of astronomers—from the stargazer who merely watches the heavens, to the abstract mathematician who merely works at his desk; it has, consequently, been necessary in the case of some lives to adopt a very different treatment from that which seemed suitable for others.

ROBERT STAWELL BALL.

The Observatory, Cambridge.
October, 1895.

NOTE: The full-length of 'Preface' can be found in its original edition of the book "GREAT ASTRONOMERS".

INTRODUCTION

Of all the natural sciences there is not one which offers such sublime objects to the attention of the inquirer as does the science of astronomy. From the earliest ages the study of the stars has exercised the same fascination as it possesses at the present day. Among the most primitive peoples, the movements of the sun, the moon, and the stars commanded attention from their supposed influence on human affairs. The practical utilities of astronomy were also obvious in primeval times. Maxims of extreme antiquity show how the avocations of the husbandman are to be guided by the movements of the heavenly bodies. The positions of the stars indicated the time to plough, and the time to sow. To the mariner who was seeking a way across the trackless ocean, the heavenly bodies .offered the only reliable marks by which his path could be guided. There was, accordingly, a stimulus both from intellectual curiosity and from practical necessity to follow the movements of the stars. Thus began a search for the causes of the ever-varying phenomena which the heavens display.

- ROBERT STAWELL BALL

Many of the earliest discoveries are indeed prehistoric. The great diurnal movement of the heavens, and the annual revolution of the sun, seem to have been known in times far more ancient than those to which any human monuments can be referred. The acuteness of the early observers enabled them to single out the more important of the wanderers which we now call planets. They saw that the star-like objects, Jupiter, Saturn, and Mars, with the more conspicuous Venus, constituted a class of bodies wholly distinct from the fixed stars among which their movements lay, and to which they bear such a superficial resemblance. But the penetration of the early astronomers went even further, for they recognized that Mercury also belongs to the same group, though this particular object is seen so rarely. It would seem that eclipses and other phenomena were observed at Babylon from a very remote period, while the most ancient records of celestial observations that we possess are to be found in the Chinese annals. The study of astronomy, in the sense in which we understand the word, may be said to have commenced under the reign of the Ptolemies at Alexandria. The most famous name in the science of this period is that of Hipparchus, who lived and worked at Rhodes about the year 160 B.C. It was his splendid investigations that first wrought the observed facts into a coherent branch of knowledge. He recognized the primary obligation which lies on the student of the heavens to compile as complete an inventory as possible of the objects which are there to be found. Hipparchus accordingly commenced by undertaking, on a small scale, a task exactly similar to that on which modern astronomers, with all available appliances of meridian circles, and photographic telescopes, are constantly engaged at the present day. He compiled a catalogue of the principal fixed stars, which is of special value to astronomers, as being the earliest work of its kind which has been handed down. He also studied the movements of the sun and the moon, and framed theories to account for the incessant changes which he saw in progress. He found a much more difficult problem in his attempt to interpret satisfactorily the complicated movements of the planets. With the view of constructing a theory which should give some coherent account of the subject, he made many observations of the places of these wandering stars. How great were the advances which Hipparchus accomplished may be appreciated if we reflect that, as a preliminary task to his more purely astronomical labours, he had to invent that branch of mathematical science by which alone the problems he proposed could be solved. It was for this purpose that he devised the indispensable method of calculation which we now know so well as trigonometry. Without the aid rendered by this beautiful art it would have been impossible for any n ally important advance in astronomical calculation to have been effected.

But the discovery which shows, beyond all others, that Hipparchus possessed one of the master-minds of all time was the detection of that remarkable celestial movement known as the precession of the equinoxes. The inquiry which conducted to this discovery involved a most profound investigation, especially when it is remembered that in the days of Hipparchus the means of observation of the heavenly bodies were only of the rudest description, and the available observations of earlier dates were extremely scanty. We can but look with astonishment on the genius of the man who, in spite of such difficulties, was able to detect such a phenomenon as the precession, and to exhibit its actual magnitude. I shall endeavour to explain the nature of this singular celestial movement, for it may be said to offer the first instance in the history of science in which we find that combination of accurate observation with skilful interpretation, of which, in the subsequent development of astronomy, we have so many splendid examples.

The word equinox implies the condition that the night is equal to the day. To a resident on the equator the night is no doubt equal to the day at all times in the year, but to one who lives on any other part of the earth, in either hemisphere, the night and the day are not generally equal. There is, however, one occasion in spring, and another in autumn, on which the day and the night are each twelve hours at all places on the earth. When the night and day are equal in spring, the point which the sun occupies on. the heavens is termed the vernal equinox. There is similarly another point in which the sun is situated at the time of the autumnal equinox. In any investigation of the celestial movements the positions of these two equinoxes on the heavens are of primary importance, and Hipparchus, with the instinct of genius, perceived their significance, and commenced to study them. It will be understood that we can always define the position of a point on the sky with reference to the surrounding stars. No doubt we do not see the stars near the sun when the sun is shining, but they are there nevertheless. The ingenuity of Hipparchus enabled him to determine the positions of each of the two equinoxes relatively to the stars which lie in its immediate vicinity. After examination of the celestial places of these points at different periods, he was led to the conclusion that each equinox was moving relatively to the stars, though that movement was so slow that twenty five thousand years would necessarily elapse before a complete circuit of the heavens was accomplished. Hipparchus traced out this phenomenon, and established it on an impregnable basis, so that all astronomers have ever since recognised the precession of the equinoxes) as one of the fundamental facts of astronomy. Not until nearly two thousand years after Hipparchus had made this splendid discovery was the explanation of its cause given by Newton.

From the days of Hipparchus down to the present hour the science of astronomy has steadily grown. One great observer after another has appeared from time to time, to reveal some new phenomenon with regard to the celestial bodies or their movements, while from time to time one commanding intellect after another has arisen to explain the true import of the facts of observations. The history of astronomy thus becomes inseparable from the history of the great men to whose labours its development is due. In the ensuing chapters we have endeavoured to sketch the lives and the work of the great philosophers, by whose labours the science of astronomy has been created.

We shall commence with Ptolemy, who, after the foundations of the science had been laid by Hipparchus, gave to astronomy the form in which it was taught throughout the Middle Ages. We shall next see the mighty revolution in our conceptions of the universe which are associated with the name of Copernicus. We then pass to those periods illumined by the genius of Galileo and Newton, and afterwards we shall trace the careers of other more recent discoverers, by whose industry and genius the boundaries of human knowledge have been so greatly extended. Our history will be brought down late enough to include some of the illustrious astronomers who laboured in the generation which has just passed away.

- ROBERT STAWELL BALL

TABLE OF CONTENTS

CHAPTERS

**An original Chapter from the book: "GREAT ASTRONOMERS", by Robert Stawell Ball.*

CHAPTER 2

GREAT ASTRONOMERS

Nicolaus Copernicus

(February 19, 1473 – May 24, 1543)

- ROBERT STAWELL BALL

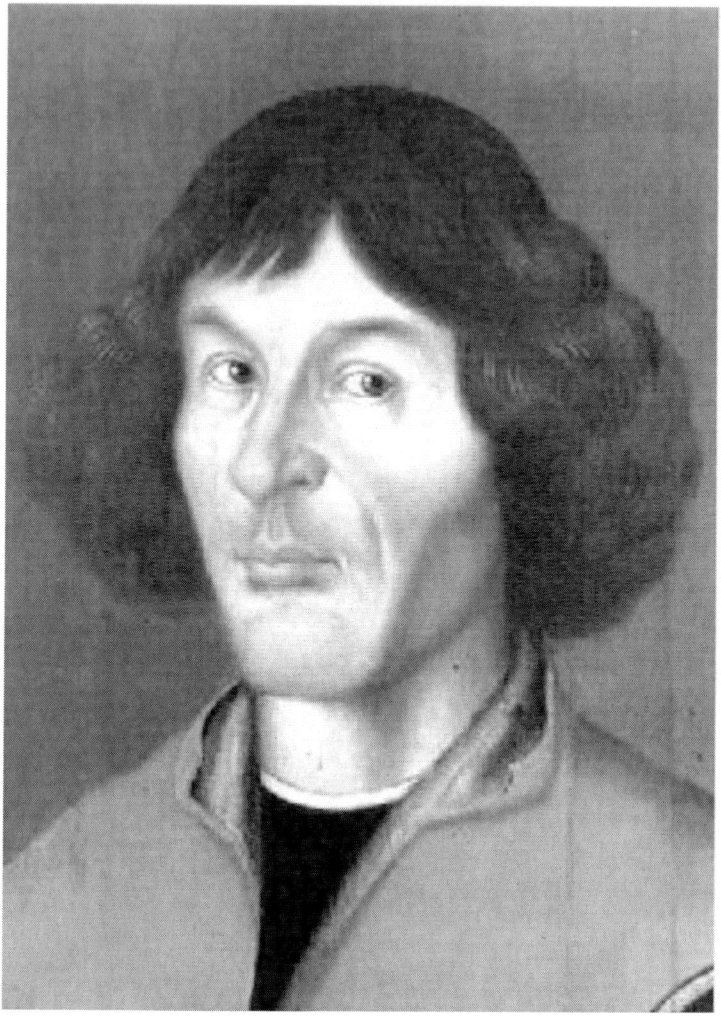

Nicolaus Copernicus

(February 19, 1473 – May 24, 1543)

CHAPTER 2

GREAT ASTRONOMERS: NICOLAUS COPERNICUS

THE quaint town of Thorn, on the Vistula, was more than two centuries old when Copernicus was born there on the 19th of February, 1473. The situation of this town on the frontier between Prussia and Poland, with the commodious waterway offered by the river, made it a place of considerable trade. A view of the town, as it was at the time of the birth of Copernicus, is here given. The walls, with their watch-towers, will be noted, and the strategic importance which the situation of Thorn gave to it in the fifteenth century still belongs thereto, so much so that the German Government recently constituted the town a fortress of the first class.

Copernicus, the astronomer, whose discoveries make him the great predecessor of Kepler and Newton, did not come from a noble family, as certain other early astronomers have done, for his father was a tradesman. Chroniclers are, however, careful to tell us that one of his uncles was a bishop. We are not acquainted with any of those details of his childhood or youth which are often of such interest in other cases where men have risen to exalted fame. It would appear that the young Nicolaus, for such was his Christian name, received his education at home until such time as he was deemed sufficiently advanced to be sent to the University at Cracow. The education that he there obtained must have been in those days of a very primitive description, but Copernicus seems to have availed himself of it to the utmost. He devoted himself more particularly to the study of medicine, with the view of adopting its practice as the profession of his life. The tendencies of the future astronomer were, however, revealed in the fact that he worked hard at mathematics, and, like one of his illustrious successors, Galileo, the practice of the art of painting had for him a very great interest, and in it he obtained some measure of success.

By the time he was twenty-seven years old, it would seem that Copernicus had given up the notion of becoming a medical practitioner, and had resolved to devote himself to science. He was engaged in teaching mathematics, and appears to have acquired some reputation. His growing fame attracted the notice of his uncle the bishop, at whose suggestion Copernicus took holy orders, and he was presently appointed to a canonry in the cathedral of Frauenhurg, near the mouth of the Vistula.

Thorn from an Old Print

To Frauenburg, accordingly, this man of varied gifts retired. Possessing somewhat of the ascetic spirit, he resolved to devote his life to work of the most serious description. He eschewed all ordinary society, restricting his intimacies to very grave and learned companions, and refusing to engage in conversation of any useless kind. It would seem as if his gifts for painting were condemned as frivolous; at all events, we do not learn that he continued to practise them. In addition to the discharge of his theological duties, his life was occupied partly in ministering medically to the wants of the poor, and partly with his researches in astronomy and mathematics. His equipment in the matter of instruments for the study of the heavens seems to have been of a very meagre description. He arranged apertures in the walls of his house at Allenstein, so that he could observe in some fashion the passage of the stars across the meridian. That he possessed some talent for practical mechanics is proved by his construction of a contrivance for raising water from a stream, for the use of the inhabitants of Frauen-burg. Relics of this machine are still to be seen.

The intellectual slumber of the Middle Ages was destined to be awakened by the revolutionary doctrines of Copernicus. It may be noted, as an interesting circumstance, that the time at which he discovered the scheme of the solar system has coincided with a remarkable epoch in the world's history. The great astronomer had just reached manhood at the time when Columbus discovered the new world.

Before the publication of the researches of Copernicus, the orthodox scientific creed averred that the earth was stationary, and that the apparent movements of the heavenly bodies were indeed real movements. Ptolemy had laid down this doctrine 1,400 years before. In his theory this huge error was associated with so much important truth, and the whole presented such a coherent scheme for the explanation of the heavenly movements, that the Ptolemaic theory was not seriously questioned until the great work of Copernicus appeared. No doubt others, before Copernicus, had from time to time in some vague fashion surmised, with more or less plausibility, that the sun, and not the earth, was the centre about which the system really revolved. It is, however, one thing to state a scientific fact; it is quite another thing to be in possession of the train of reasoning, founded on observation or experiment, by which that fact may be established. Pythagoras, it appears, had indeed told his disciples that it was the sun, and not the earth, which was the centre of movement, but it does not seem at all certain that Pythagoras had any grounds which science could recognise for the belief which is attributed to him. So far as information is available to us, it would seem that Pythagoras associated his scheme of things celestial with a number of preposterous notions in natural philosophy. He may certainly have made a correct statement as to which was the most important body in the solar system, but he certainly did not provide any rational demonstration of the fact. Copernicus, by a strict train of reasoning, convinced those who would listen to him that the sun was the centre of the system. It is useful for us to consider the arguments which he urged, and by which he effected that intellectual revolution which is always connected with his name.

The first of the great discoveries which Copernicus made relates to the rotation of the earth on its axis. That general diurnal movement, by which the stars and all other celestial bodies appear to be carried completely round the heavens once every twenty-four hours, had been accounted for by Ptolemy on the supposition that the apparent movements were the real movements. As we have already seen, Ptolemy himself felt the extraordinary difficulty involved in the supposition that so stupendous a fabric as the celestial sphere should spin in the way supposed. Such movements required that many of the stars should travel with almost inconceivable velocity. Copernicus also saw that the daily rising and setting of the heavenly bodies could be accounted for either by the supposition that the celestial sphere moved round and that the earth remained at rest, or by the supposition that the celestial sphere was at rest while the earth turned round in the opposite direction. He weighed the arguments on both sides as Ptolemy had done, and, as the result of his deliberations, Copernicus came to an opposite conclusion from Ptolemy. To Copernicus it appeared that the difficulties attending the supposition that the celestial sphere revolved, were vastly greater than those which appeared so weighty to Ptolemy as to force him to deny the earth's rotation.

Copernicus shows clearly how the observed phenomena could be accounted for just as completely by a rotation of the earth as by a rotation of the heavens. He alludes to the fact that, to those on board a vessel which is moving through smooth water, the vessel itself appears to be at rest, while the objects on shore seem to be moving past. If, therefore, the earth were rotating uniformly, we dwellers upon the earth, oblivious of our own movement, would wrongly attribute to the stars the displacement which was actually the consequence of our own motion.

Copernicus saw the futility of the arguments by which Ptolemy had endeavoured to demonstrate that a revolution of the earth was impossible. It was plain to him that there was nothing whatever to warrant refusal to believe in the rotation of the earth. In his clear-sightedness on this matter we have specially to admire the sagacity of Copernicus as a natural philosopher. It had been urged that, if the earth moved round, its motion would not be imparted to the air, and that therefore the earth would be uninhabitable by the terrific winds which would be the result of our being carried through the air. Copernicus convinced himself that this deduction was preposterous. He proved that the air must accompany the earth, just as his coat remains round him, notwithstanding the fact that he is walking down the street. In this way he was able to show that all a *priori* objections to the earth's movements were absurd, and therefore he was able to compare together the plausibilities of the two rival schemes for explaining the diurnal movement.

Once the issue had been placed in this form, the result could not be long in doubt. Here is the question: Which is it more likely—that the earth, like a grain of sand at the centre of a mighty globe, should turn round once in twenty-four hours, or—that the whole of that vast globe should complete a rotation in the opposite direction in the same time? Obviously, the former is far the more simple supposition. But the case is really much stronger than this. Ptolemy had supposed that all the stars were attached to the surface of a sphere. He had no ground whatever for this supposition, except that otherwise it would have been well-nigh impossible to have devised a scheme by which the rotation of the heavens around a fixed earth could have been arranged. Copernicus, however, with the just instinct of a philosopher, considered that the celestial sphere, however convenient from a geometrical point of view, as a means of representing apparent phenomena, could not actually have a material existence.

In the first place, the existence of a material celestial sphere would require that all the myriad stars should be at exactly the same distances from the earth.

Of course, no one will say that this or any other arbitrary disposition of the stars is actually *impossible*, but as there was no

Frauenburg from an Old Print

- ROBERT STAWELL BALL

conceivable physical reason why the distances of all the stars from the earth should be identical, it seemed in the very highest degree improbable that the stars should be so placed.

Doubtless, also, Copernicus felt a considerable difficulty as to the nature of the materials from which Ptolemy's wonderful sphere was to be constructed. Nor could a philosopher of his penetration have failed to observe that, unless that sphere were infinitely large, there must have been space outside it, a consideration which would open up other difficult questions. Whether infinite or not, it was obvious that the celestial sphere must have a diameter at least many thousands of times as great as that of the earth. From these considerations Copernicus deduced the important fact that the stars and the other celestial bodies must all be vast objects. He was thus enabled to put the question in such a form that it could hardly receive any answer but the correct one. Which is it more rational to suppose, that the earth should turn round on its axis once in twenty-four hours, or that thousands of mighty stars should circle round the earth in the same time, many of them having to describe circles many thousands of times greater in circumference than the circuit of the earth at the equator? The obvious answer pressed upon Copernicus with so much force that he was compelled to reject Ptolemy's theory of the stationary earth, and to attribute the diurnal rotation of the heavens to the revolution of the earth on its axis.

Once this tremendous step had been taken, the great difficulties which beset the monstrous conception of the celestial sphere vanished, for the stars need no longer be regarded as situated at equal distances from the earth. Copernicus saw that they might lie at the most varied degrees of remoteness, some being hundreds or thousands of times farther away than others. The complicated structure of the celestial sphere as a material object disappeared altogether; it remained only as a geometrical conception, whereon we find it convenient to indicate the places of the stars. Once the Copernican doctrine had been fully set forth, it was impossible for anyone, who had both the inclination and the capacity to understand it, to withhold acceptance of its truth. The doctrine of a stationary earth had gone forever.

Copernicus having established a theory of the celestial movements which deliberately set aside the stability of the earth, it seemed natural that he should inquire whether the doctrine of a moving earth might not remove the difficulties presented in other celestial phenomena. It had been universally admitted that the earth lay unsupported in space. Copernicus had further shown that it possessed a movement of rotation. Its want of stability being thus recognised, it seemed reasonable to suppose that the earth might also have some other kinds of movements as well. In this, Copernicus essayed to solve a problem far more difficult than that which had hitherto occupied his attention. It was a comparatively easy task to show how the diurnal rising and setting could be accounted for by the rotation, of the earth. It was a much more difficult undertaking to demonstrate that the planetary movements, which Ptolemy had represented with so much success, could be completely explained by the supposition that each of those planets revolved uniformly round the sun, and that the earth was also a planet, accomplishing a complete circuit of the sun once in the course of a year.

It would be impossible in a sketch like the present to enter into any detail as to the geometrical propositions on which this beautiful investigation of Copernicus depended. We can only mention a few of the leading principles. It may be laid down in general that, if an observer is in movement, he will, if unconscious of the fact, attribute to the fixed objects around him a movement equal and opposite to that which he actually possesses. A passenger on a canal-boat sees the objects on the banks apparently moving backward with a speed equal to that by which he is himself advancing forwards. By an application of this principle, we can account for all the phenomena of the movements of the planets, which Ptolemy had so ingeniously

represented by his circles. Let us take, for instance, the most characteristic feature in. the irregularities of the outer planets. We have already remarked that Mars, though generally advancing from west to east among the stars, occasionally pauses, retraces his steps for a while, again pauses, and then resumes his ordinary onward progress. Copernicus showed clearly how this effect was produced by the real motion of the earth, combined with the real motion of Mars. In the adjoining figure we represent a portion of the circular tracks in which the earth and Mars move in accordance with the Copernican doctrine. I show particularly the case where the earth comes directly between the planet and the sun, because it is on such occasions that the retrograde movement (for so this backward movement of Mars is termed) is at its highest. Mars is then advancing in the direction shown by the arrow-head, and the earth is also advancing in the same direction. We, on the earth, however, being unconscious of our own motion, attribute, by the principle I have already explained, an equal and opposite motion to Mars.

The visible effect upon the planet is, that Mars has two movements, a real onward movement in one direction, and an apparent movement in the opposite direction. If it so happened that the earth was moving with the same speed as Mars,

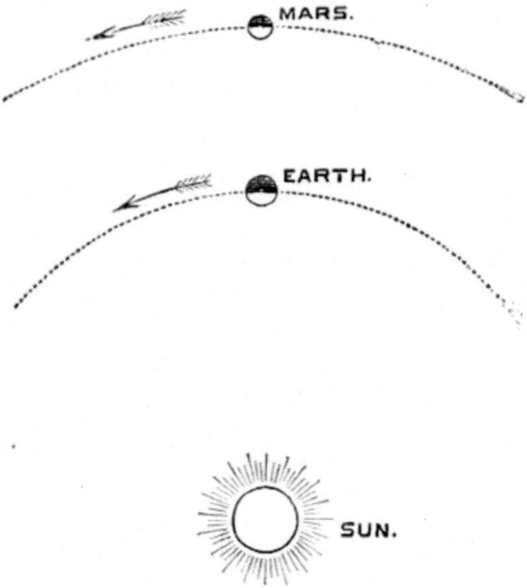

then the apparent movement would exactly neutralise the real movement, and Mars would seem to be at rest relatively to the surrounding stars. Under the actual circumstances represented, however, the earth is moving faster than Mars, and the consequence is, that the apparent movement of the planet backwards exceeds the real movement forwards, the net result being an apparent retrograde movement.

With consummate skill, Copernicus showed how the applications of the same principles could account for the characteristic movements of the planets. His reasoning in due time bore down all opposition. The supreme importance of the earth in the system vanished. It had now merely to take rank as one of the planets.

The same great astronomer now, for the first time, rendered something like a rational account of the changes of the seasons. Nor did certain of the more obscure astronomical phenomena escape his attention.

He delayed publishing his wonderful discoveries to the world until he was quite an old man. He had a well-founded apprehension of the storm of opposition which they would arouse.3 However, he yielded at last to the entreaties of his friends, and his book was sent to the press. But ere it made its appearance to the world, Copernicus was seized by mortal illness. A copy of the book was brought to him on May 23, 1543. We are told that he was able to see it and to touch it, but no more, and he died a few hours afterwards. He was buried in that Cathedral of Frauenburg, with which his life had been so closely associated.

COMPLIMENTARY MATERIAL

A BRIEF BIOGRAPHICAL SKETCH

NICOLAUS COPERNICUS

- ROBERT STAWELL BALL

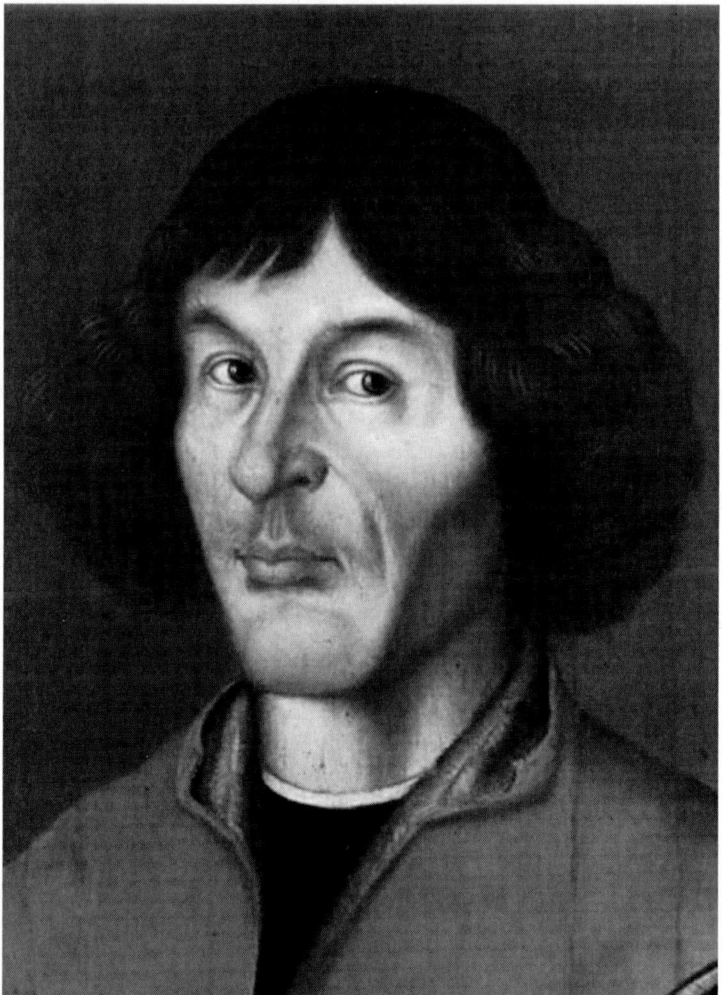

Nicolaus Copernicus
(February 19, 1473 – May 24, 1543)

COMPLIMENTARY MATERIAL

A BRIEF BIOGRAPHICAL SKETCH

NICOLAUS COPERNICUS

Nicolaus Copernicus (19 February 1473 – 24 May 1543) was a Renaissance mathematician and astronomer who formulated a model of the universe that placed the Sun rather than the Earth at the center of the universe. The publication of this model in his book *De revolutionibus orbium coelestium* (On the Revolutions of the Celestial Spheres) just before his death in 1543 is considered a major event in the history of science, triggering the Copernican Revolution and making an important contribution to the Scientific Revolution.

Copernicus was born and died in Royal Prussia, a region that had been a part of the Kingdom of Poland since 1466. He was a polyglot and polymath who obtained a doctorate in canon law and also practiced as a physician, classics scholar, translator, governor, diplomat, and economist. Like the rest of his family, he was a third-order Dominican. In 1517 he derived a quantity theory of money – a key concept in economics – and in 1519 he formulated a version of what later became known as Gresham's law.

Nicolaus Copernicus was born on 19 February 1473 in the city of Toruń (Thorn), in the province of Royal Prussia, in the Crown of the Kingdom of Poland. His father was a merchant from Kraków and his mother was the daughter of a wealthy Toruń merchant. Nicolaus was the youngest of four children. His brother Andreas (Andrew) became an Augustinian canon at Frombork (Frauenburg). His sister Barbara, named after her mother,

Toruń birthplace (ul. Kopernika 15, left). Together with the house at no. 17 (right), it forms the Muzeum Mikołaja Kopernika.

- ROBERT STAWELL BALL

became a Benedictine nun and, in her final years, prioress of a convent in Chełmno (Kulm); she died after 1517. His sister Katharina married the businessman and Toruń city councilor Barthel Gertner and left five children, whom Copernicus looked after to the end of his life. Copernicus never married or had children.

Father's family

The father's family can be traced to a village in Silesia near Nysa (Neiße). The village's name has been variously spelled Kopernik, Copernik, Copernic, Kopernic, Coprirnik, and today Koperniki. In the 14th century, members of the family began moving to various other Silesian cities, to the Polish capital, Kraków (1367), and to Toruń (1400). The father, Mikołaj the Elder, likely the son of Jan, came from the Kraków line.

Nicolaus was named after his father, who appears in records for the first time as a well-to-do merchant who dealt in copper, selling it mostly in Danzig (Gdańsk). He moved from Kraków to Toruń around 1458. Toruń, situated on the Vistula River, was at that time embroiled in the Thirteen Years' War, in which the Kingdom of Poland and the Prussian Confederation, an alliance of Prussian cities, gentry and clergy, fought the Teutonic Order over control of the region. In this war, Hanseatic cities like Danzig and Toruń, Nicolaus Copernicus's hometown, chose to support the Polish King, Casimir IV Jagiellon, who promised to respect the cities' traditional vast independence, which the Teutonic Order had challenged. Nicolaus' father was actively engaged in the politics of the day and supported Poland and the cities against the Teutonic Order. In 1454 he mediated negotiations between Poland's Cardinal Zbigniew Oleśnicki and the Prussian cities for repayment of war loans. In the Second Peace of Thorn (1466), the Teutonic Order formally relinquished all claims to its western province, which as Royal Prussia remained a region of the Crown of the Kingdom of Poland until the First (1772) and Second (1793) Partitions of Poland. Copernicus's father married Barbara Watzenrode, the astronomer's mother, between 1461 and 1464. He died about 1483.

Mother's family

Nicolaus' mother, Barbara Watzenrode, was the daughter of a wealthy Toruń patrician and city councillor, Lucas Watzenrode the Elder (deceased 1462), and Katarzyna (widow of Jan Peckau), mentioned in other sources as Katarzyna Rüdiger gente Modlibóg (deceased 1476). The Modlibógs were a prominent Polish family who had been well known in Poland's history since 1271. The Watzenrode family, like the Kopernik family, had come from Silesia from near Świdnica (Schweidnitz), and after 1360 had settled in Toruń. They soon became one of the wealthiest and most influential patrician families. Through the Watzenrodes' extensive family relationships by marriage, Copernicus was related to wealthy families of Toruń (Thorn), Gdańsk (Danzig) and Elbląg (Elbing), and to prominent Polish noble families of Prussia: the Czapskis, Działyńskis, Konopackis and Kościeleckis. Lucas and Katherine had three children: Lucas Watzenrode the Younger (1447–1512), who would become Bishop of Warmia and Copernicus's patron; Barbara, the astronomer's mother (deceased after 1495); and Christina (deceased before 1502), who in 1459 married the Toruń merchant and mayor, Tiedeman von Allen.

Copernicus' maternal uncle,
Lucas Watzenrode the Younger

Lucas Watzenrode the Elder, a wealthy merchant and in 1439–62 president of the judicial bench, was a decided opponent of the Teutonic Knights. In 1453 he was the delegate from Toruń at the Grudziądz (Graudenz) conference that planned the uprising against them. During the ensuing Thirteen Years' War (1454–66), he actively supported the Prussian cities' war effort with substantial monetary subsidies (only part of which he later re-claimed), with political activity in Toruń and Danzig, and by personally fighting in battles at Łasin (Lessen) and Malbork (Marienburg).

He died in 1462.

Lucas Watzenrode the Younger, the astronomer's maternal uncle and patron, was educated at the University of Kraków (now Jagiellonian University) and at the universities of Cologne and Bologna. He was a bitter opponent of the Teutonic Order, and its Grand Master once referred to him as "the devil incarnate". In 1489

Watzenrode was elected Bishop of Warmia (Ermeland, Ermland) against the preference of King Casimir IV, who had hoped to install his own son in that seat. As a result, Watzenrode quarreled with the king until Casimir IV's death three years later. Watzenrode was then able to form close relations with three successive Polish monarchs: John I Albert, Alexander Jagiellon, and Sigismund I the Old. He was a friend and key advisor to each ruler, and his influence greatly strengthened the ties between Warmia and Poland proper. Watzenrode came to be considered the most powerful man in Warmia, and his wealth, connections and influence allowed him to secure Copernicus' education and career as a canon at Frombork Cathedral.

Languages

Copernicus is postulated to have spoken Latin and German with equal fluency. He also spoke Polish, Greek and Italian. The vast majority of Copernicus's surviving works are in Latin, which in his lifetime was the language of academia in Europe. Latin was also the official language of the Roman Catholic Church and of Poland's royal court, and thus all of Copernicus's correspondence with the Church and with Polish leaders was in Latin.

There survive a few documents written by Copernicus in German. The German philosophy professor Martin Carrier mentions this as a reason to consider Copernicus's native language to have been German. Other arguments for German being Copernicus's native tongue are that he was born in a predominantly German-speaking city and that, while studying canon law at Bologna in 1496, he signed into the German natio (Natio Germanorum)—a student organization which, according to its 1497 by-laws, was open to students of all kingdoms and states whose mother-tongue was German. However, according to French philosopher Alexandre Koyré, Copernicus's registration with the Natio Germanorum does not in itself imply that Copernicus considered himself German, since students from Prussia and Silesia were routinely so categorized, which carried certain privileges that made it a natural choice for German-speaking students, regardless of their ethnicity or self-identification.

Tafel III.

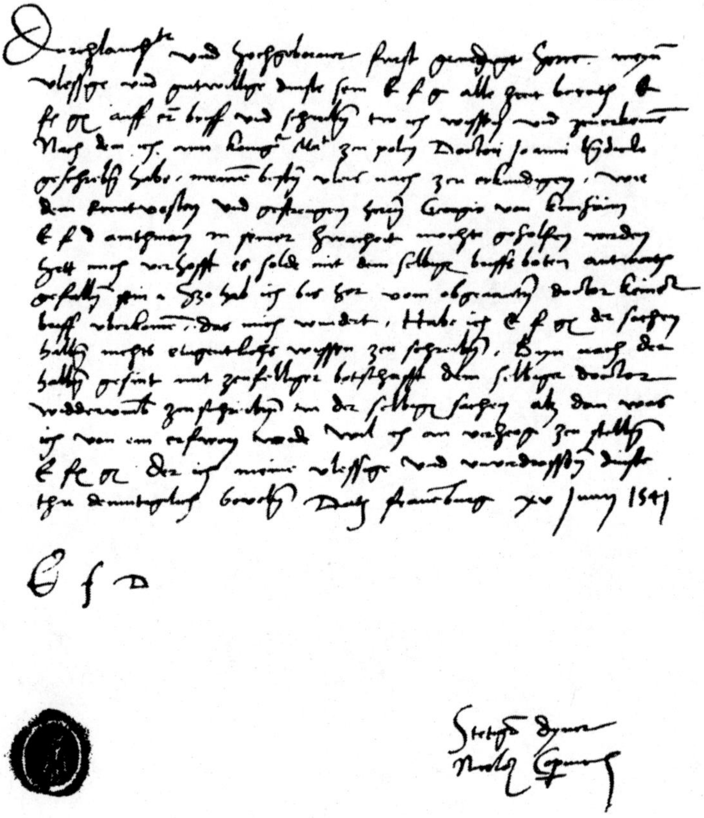

German-language letter from Copernicus to Duke Albert of Prussia, giving medical advice for George von Kunheim

- ROBERT STAWELL BALL

Name

In Copernicus's time, people were often called after the places where they lived. Like the Silesian village that inspired it, Copernicus's surname has been spelled variously. The surname likely had something to do with the local Silesian copper-mining industry, though some scholars assert that it may have been inspired by the dill plant (in Polish, "*koperek*" or "*kopernik*") that grows wild in Silesia.

As was to be the case with William Shakespeare a century later, numerous variant spellings of the name are documented for the astronomer and his relatives. The name first appeared as a place name in Silesia in the 13th century, where it was spelled variously in Latin documents. Copernicus "was rather indifferent about orthography". During his childhood, about 1480, the name of his father (and thus of the future astronomer) was recorded in Thorn as *Niclas Koppernigk*. At Kraków he signed himself, in Latin, *Nicolaus Nicolai de Torunia* (Nicolaus, son of Nicolaus, of Toruń). At Bologna, in 1496, he registered in the *Matricula Nobilissimi Germanorum Collegii, resp. Annales Clarissimae Nacionis Germanorum*, of the *Natio Germanica Bononiae, as Dominus Nicolaus Kopperlingk de Thorn* – IX *grosseti*. At Padua he signed himself "Nicolaus Copernik", later "Coppernicus". The astronomer thus Latinized his name to Coppernicus, generally with two "p"s (in 23 of 31 documents studied), but later in life he used a single "p". On the title page of *De revolutionibus*, Rheticus published the name as (in the genitive, or possessive, case) "Nicolai Copernici".

- ROBERT STAWELL BALL

Education

In Poland

Upon his father's death, young Nicolaus' maternal uncle, Lucas Watzenrode the Younger (1447–1512), took the boy under his wing and saw to his education and career. Watzenrode maintained contacts with leading intellectual figures in Poland and was a friend of the influential Italian-born humanist and Kraków courtier, Filippo Buonaccorsi. There are no surviving primary documents on the early years of Copernicus's childhood and education. Copernicus biographers assume that Watzenrode first sent young Copernicus to St. John's School, at Toruń, where he himself had been a master. Later, according to Armitage, the boy attended the Cathedral School at Włocławek, up the Vistula River from Toruń, which prepared pupils for entrance to the University of Kraków, Watzenrode's alma mater in Poland's capital.

The Collegium Maius
(Latin for "Great College"), Kraków

In the winter semester of 1491–92 Copernicus, as "Nicolaus Nicolai de Thuronia", matriculated together with his brother Andrew at the University of Kraków (now Jagiellonian University). Copernicus began his studies in the Department of Arts (from the fall of 1491,

Nicolaus Copernicus Monument in Kraków

presumably until the summer or fall of 1495) in the heyday of the Kraków astronomical-mathematical school, acquiring the foundations for his subsequent mathematical achievements. According to a later but credible tradition (Jan Brożek), Copernicus was a pupil of Albert Brudzewski, who by then (from 1491) was a professor of Aristotelian philosophy but taught astronomy privately outside the university; Copernicus became familiar with Brudzewski's widely read commentary to Georg von Peuerbach's Theoricæ novæ planetarum and almost certainly attended the lectures of Bernard of Biskupie and Wojciech Krypa of Szamotuły, and probably other astronomical lectures by Jan of Głogów, Michał of Wrocław (Breslau), Wojciech of Pniewy, and Marcin Bylica of Olkusz.

Copernicus's Kraków studies gave him a thorough grounding in the mathematical-astronomical knowledge taught at the university (arithmetic, geometry, geometric optics, cosmography, theoretical and computational astronomy), a good knowledge of the philosophical and natural-science writings of Aristotle (De coelo, Metaphysics), stimulated his interest in learning, and made him conversant with humanistic culture. Copernicus broadened the knowledge that he took from the university lecture halls with independent reading of books that he acquired during his Kraków years (Euclid, Haly Abenragel, the Alfonsine Tables, Johannes Regiomontanus' Tabulae directionum); to this period, probably, also date his earliest scientific notes, now preserved partly at Uppsala University. At Kraków Copernicus began collecting a large library on astronomy; it would later be carried off as war booty by the Swedes during the Deluge in the 1650s and is now at the Uppsala University Library.

Copernicus's four years at Kraków played an important role in the development of his critical faculties and initiated his analysis of the logical contradictions in the two most popular systems of astronomy—Aristotle's theory of homocentric spheres, and Ptolemy's mechanism of eccentrics and epicycles—the surmounting and discarding of which constituted the first step toward the creation of Copernicus's own doctrine of the structure of the universe.

Without taking a degree, probably in the fall of 1495, Copernicus left Kraków for the court of his uncle Watzenrode, who in 1489 had been elevated to Prince-Bishop of Warmia and soon (before November 1495) sought to place his nephew in the Warmia canonry vacated by 26 August 1495 death of its previous tenant, Jan Czanow. For unclear reasons—probably due to opposition from part of the chapter, who appealed to Rome—Copernicus's installation was delayed, inclining Watzenrode to send both his nephews to study canon law in Italy, seemingly with a view to furthering their ecclesiastic careers and thereby also strengthening his own influence in the Warmia chapter.

- ROBERT STAWELL BALL

Leaving Warmia in mid-1496—possibly with the retinue of the chapter's chancellor, Jerzy Pranghe, who was going to Italy—in the fall, possibly the October, of that year Copernicus arrived in Bologna and a few months later (after 6 January 1497) signed himself into the register of the Bologna University of Jurists' "German nation", which also included young Poles from Silesia, Prussia and Pomerania, as well as students of other nationalities.

In Italy

It was only on 20 October 1497 that Copernicus, by proxy, formally succeeded to the Warmia canonry which had been granted to him two years earlier. To this, by a document dated 10 January 1503 at Padua, he would add a sinecure at the Collegiate Church of the Holy Cross in Wrocław, Silesia, Bohemia. Despite having been granted a papal indult on 29 November 1508 to receive further benefices, through his ecclesiastic career Copernicus not only did not acquire further prebends and higher stations (prelacies) at the chapter, but in 1538 he relinquished the Wrocław sinecure. It is unclear whether he was ever ordained a priest. Author Ed Rosen asserts that he was not. He did take minor orders, which sufficed for assuming a chapter canonry. The Catholic encyclopedia proposes that his ordination was probable, as he was one of four candidates for the episcopal seat of Ermland, a position which required ordination. During his three-year stay at Bologna, between fall 1496 and spring 1501, Copernicus seems to have devoted himself less keenly to studying canon law (he received his doctorate in law only after seven years, following a second return to Italy in 1503) than to studying the humanities—probably attending lectures by Filippo Beroaldo, Antonio Urceo, called Codro, Giovanni Garzoni, and Alessandro Achillini—and to studying astronomy. He met the famous astronomer Domenico Maria Novara da Ferrara and became his disciple and assistant. Copernicus was developing new ideas inspired by reading the "Epitome of the *Almagest*" (*Epitome in Almagestum Ptolemei*) by George von Peuerbach and Johannes Regiomontanus (Venice, 1496). He verified its observations about certain peculiarities in Ptolemy's theory of the Moon's motion, by conducting on 9 March 1497 at Bologna a memorable observation of the occultation of Aldebaran, the brightest star in the Taurus constellation, by the moon.

Copernicus the humanist sought confirmation for his growing doubts through close reading of Greek and Latin authors (Pythagoras, Aristarchos of Samos, Cleomedes, Cicero, Pliny the Elder, Plutarch, Philolaus, Heraclides, Ecphantos, Plato), gathering, especially while at Padua, fragmentary historic information about ancient astronomical, cosmological and calendar systems.

Copernicus spent the jubilee year 1500 in Rome, where he arrived with his brother Andrew that spring, doubtless to perform an apprenticeship at the Papal Curia. Here, too, however, he continued his astronomical work begun at Bologna, observing, for example, a lunar eclipse on the night of 5–6 November 1500. According to a later account by Rheticus, Copernicus also—probably privately, rather than at the Roman *Sapienza*—as a "Professor Mathematatum" (professor of astronomy) delivered, "to numerous... students and... leading masters of the science", public lectures devoted probably to a critique of the mathematical solutions of contemporary astronomy.

Via Galliera 65, Bologna, site of house of
Domenico Maria Novara.
Plaque on portico commemorates Copernicus.

"Here, where stood the house of Domenico Maria Novara, professor of the ancient Studium of Bologna, NICOLAUS COPERNICUS, the Polish mathematician and astronomer who would revolutionize concepts of the universe, conducted brilliant celestial observations with his teacher in 1497–1500. Placed on the 5th centenary of [Copernicus's] birth by the City, the University, the Academy of Sciences of the Institute of Bologna, the Polish Academy of Sciences. 1473 [—] 1973."

On his return journey doubtless stopping briefly at Bologna, in mid-1501 Copernicus arrived back in Warmia. After on 28 July receiving from the chapter a two-year extension of leave in order to study medicine (since "he may in future be a useful medical advisor to our Reverend Superior [Bishop Lucas Watzenrode] and the gentlemen of the chapter"), in late summer or in the fall he returned again to Italy, probably accompanied by his brother Andrew and by Canon B. Sculteti. This time he studied at the University of Padua, famous as a seat of medical learning, and—except for a brief visit to Ferrara in May–June 1503 to pass examinations for, and receive, his doctorate in canon law—he remained at Padua from fall 1501 to summer 1503.

Copernicus studied medicine probably under the direction of leading Padua professors—Bartolomeo da Montagnana, Girolamo Fracastoro, Gabriele Zerbi, Alessandro Benedetti—and read medical treatises that he acquired at this time, by Valescus de Taranta, Jan Mesue, Hugo Senensis, Jan Ketham, Arnold de Villa Nova, and Michele Savonarola, which would form the embryo of his later medical library.

One of the subjects that Copernicus must have studied was astrology, since it was considered an important part of a medical education. However, unlike most other prominent Renaissance astronomers, he appears never to have practiced or expressed any interest in astrology.

As at Bologna, Copernicus did not limit himself to his official studies. It was probably the Padua years that saw the beginning of his Hellenistic interests. He familiarized himself with Greek language and culture with the aid of Theodorus Gaza's grammar (1495) and J.B. Chrestonius' dictionary (1499), expanding his studies of antiquity, begun at Bologna, to the writings of Basilius Bessarion, Lorenzo Valla and others. There also seems to be evidence that it was during his Padua stay that the idea finally crystallized, of basing a new system of the world on the movement of the Earth. As the time approached for Copernicus to return home, in spring 1503 he journeyed to Ferrara where, on 31 May 1503, having passed the obligatory examinations, he was granted the degree of doctor of canon law (*Nicolaus Copernich de Prusia, Jure Canonico ... et doctoratus*). No doubt it was soon after (at latest, in fall 1503) that he left Italy for good to return to Warmia.

Work

Having completed all his studies in Italy, 30-year-old Copernicus returned to Warmia, where he would live out the remaining 40 years of his life, apart from brief journeys to Kraków and to nearby Prussian cities: Toruń (Thorn), Gdańsk (Danzig), Elbląg (Elbing), Grudziądz (Graudenz), Malbork (Marienburg), Königsberg (Królewiec).

The Prince-Bishopric of Warmia enjoyed substantial autonomy, with its own diet (parliament) and monetary unit (the same as in the other parts of Royal Prussia) and treasury.

Copernicus was his uncle's secretary and physician from 1503 to 1510 (or perhaps till his uncle's death on 29 March 1512) and resided in the Bishop's castle at Lidzbark (Heilsberg), where he began work on his heliocentric theory. In his official capacity, he took part in nearly all his uncle's political, ecclesiastic and administrative-economic duties. From the beginning of 1504, Copernicus accompanied Watzenrode to sessions of the Royal Prussian diet held at Malbork and Elbląg and, write Dobrzycki and Hajdukiewicz, "participated... in all the more important events in the complex diplomatic game that ambitious politician and statesman played in defense of the particular interests of Prussia and Warmia, between hostility to the [Teutonic] Order and loyalty to the Polish Crown."

In 1504–12 Copernicus made numerous journeys as part of his uncle's retinue—in 1504, to Toruń and Gdańsk, to a session of the Royal Prussian Council in the presence of Poland's King Alexander Jagiellon; to sessions of the Prussian diet at Malbork (1506), Elbląg (1507) and Sztum (Stuhm) (1512); and he may have attended a Poznań (Posen) session (1510) and the

Copernicus's translation of Theophylact Simocatta's Epistles. Cover shows coats-of-arms of (clockwise from top) Poland, Lithuania and Kraków.

coronation of Poland's King Sigismund I the Old in Kraków (1507). Watzenrode's itinerary suggests that in spring 1509 Copernicus may have attended the Kraków sejm.

It was probably on the latter occasion, in Kraków, that Copernicus submitted for printing at Jan Haller's press his translation, from Greek to Latin, of a collection, by the 7th-century Byzantine historian Theophylact Simocatta, of 85 brief poems called Epistles, or letters, supposed to have passed between various characters in a Greek story. They are of three kinds—"moral," offering advice on how people should live; "pastoral", giving little pictures of shepherd life; and "amorous", comprising love poems. They are arranged to follow one another in a regular rotation of subjects. Copernicus had translated the Greek verses into Latin prose, and he now published his version as *Theophilacti scolastici Simocati epistolae morales, rurales et amatoriae interpretatione latina*, which he dedicated to his uncle in gratitude for all the benefits he had received from him. With this translation, Copernicus declared himself on the side of the humanists in the struggle over the question whether Greek literature should be revived. Copernicus's first poetic work was a Greek epigram, composed probably during a visit to Kraków, for Johannes Dantiscus' epithalamium for Barbara Zapolya's 1512 wedding to King Zygmunt I the Old.

Some time before 1514, Copernicus wrote an initial outline of his heliocentric theory known only from later transcripts, by the title (perhaps given to it by a copyist), *Nicolai Copernici de hypothesibus motuum coelestium a se constitutis commentariolus*—commonly referred to as the *Commentariolus*. It was a succinct theoretical description of the world's heliocentric mechanism, without mathematical apparatus, and differed in some important details of geometric construction from *De revolutionibus*; but it was already based on the same assumptions regarding Earth's triple motions. The *Commentariolus*, which Copernicus consciously saw as merely a first sketch for his planned book, was not intended for printed distribution. He made only a very few manuscript copies available to his closest acquaintances, including, it seems, several Kraków astronomers with whom he collaborated in 1515–30 in observing eclipses. Tycho Brahe would include a fragment from the *Commentariolus* in his own treatise, *Astronomiae instauratae progymnasmata*, published in Prague in 1602, based on a manuscript that he had received from the Bohemian physician and astronomer Tadeáš Hájek, a friend of Rheticus. The *Commentariolus* would appear complete in print for the first time only in 1878.

In 1510 or 1512 Copernicus moved to Frombork, a town to the northwest at the Vistula Lagoon on the Baltic Sea coast.

There, in April 1512, he participated in the election of Fabian of Lossainen as Prince-Bishop of Warmia. It was only in early June 1512 that the chapter gave Copernicus an "external curia"—a house outside the defensive walls of the cathedral mount. In 1514 he purchased the northwestern tower within the walls of the Frombork stronghold. He would maintain both these residences to the end of his life, despite the devastation of the chapter's buildings by a raid against Frauenburg carried out by the Teutonic Order in January 1520, during which Copernicus's astronomical instruments were probably destroyed. Copernicus conducted astronomical observations in 1513–16 presumably from his external curia; and in 1522–43, from an unidentified "small tower" (*turricula*), using primitive instruments modeled on ancient ones—the quadrant, triquetrum, armillary sphere. At Frombork Copernicus conducted over half of his more than 60 registered astronomical observations.

Copernicus's tower at Frombork, where he lived and worked; rebuilt recently

- ROBERT STAWELL BALL

Having settled permanently at Frombork, where he would reside to the end of his life, with interruptions in 1516–19 and 1520–21, Copernicus found himself at the Warmia chapter's economic and administrative center, which was also one of Warmia's two chief centers of political life. In the difficult, politically complex situation of Warmia, threatened externally by the Teutonic Order's aggressions (attacks by Teutonic bands; the Polish-Teutonic War of 1519–21; Albert's plans to annex Warmia), internally subject to strong separatist pressures (the selection of the prince-bishops of Warmia; currency reform), he, together with part of the chapter, represented a program of strict cooperation with the Polish Crown and demonstrated in all his public activities (the defense of his country against the Order's plans of conquest; proposals to unify its monetary system with the Polish Crown's; support for Poland's interests in the Warmia dominion's ecclesiastic administration) that he was consciously a citizen of the Polish-Lithuanian Republic. Soon after the death of uncle Bishop Watzenrode, he participated in the signing of the Second Treaty of Piotrków Trybunalski (7 December 1512), governing the appointment of the Bishop of Warmia, declaring, despite opposition from part of the chapter, for loyal cooperation with the Polish Crown.

Frombork Cathedral mount and fortifications.
In foreground: statue of Copernicus

That same year (before 8 November 1512) Copernicus assumed responsibility, as magister pistoriae, for administering the chapter's economic enterprises (he would hold this office again in 1530), having already since 1511 fulfilled the duties of chancellor and visitor of the chapter's estates.

His administrative and economic dutes did not distract Copernicus, in 1512–15, from intensive observational activity. The results of his observations of Mars and Saturn in this period, and especially a series of four observations of the Sun made in 1515, led to discovery of the variability of Earth's eccentricity and of the movement of the solar apogee in relation to the fixed stars, which in 1515–19 prompted his first revisions of certain assumptions of his system. Some of the observations that he made in this period may have had a connection with a proposed reform of the Julian calendar made in the first half of 1513 at the request of the Bishop of Fossombrone, Paul of Middelburg. Their contacts in this matter in the period of the Fifth Lateran Council were later memorialized in a complimentary mention in Copernicus's dedicatory epistle in *Dē revolutionibus orbium coelestium* and in a treatise by Paul of Middelburg, *Secundum compendium correctionis Calendarii* (1516), which mentions Copernicus among the learned men who had sent the Council proposals for the calendar's emendation.

During 1516–21, Copernicus resided at Olsztyn (Allenstein) Castle as economic administrator of Warmia, including Olsztyn (Allenstein) and Pieniężno (Mehlsack). While there, he wrote a manuscript, *Locationes mansorum desertorum* (Locations of Deserted Fiefs), with a view to populating those fiefs with industrious farmers and so bolstering the economy of Warmia. When Olsztyn was besieged by the Teutonic Knights during the Polish–Teutonic War, Copernicus directed the defense of Olsztyn and Warmia by Royal Polish forces. He also represented the Polish side in the ensuing peace negotiations.

Olsztyn Castle

Copernicus for years advised the Royal Prussian sejmik on monetary reform, particularly in the 1520s when that was a major question in regional Prussian politics. In 1526 he wrote a study on the value of money, *Monetae cudendae ratio*. In it he formulated an early iteration of the theory, now called Gresham's law, that "bad" (debased) coinage drives "good" (undebased) coinage out of circulation—several decades before Thomas Gresham. He also, in 1517, set down a quantity theory of money, a principal concept in economics to the present day. Copernicus's recommendations on monetary reform were widely read by leaders of both Prussia and Poland in their attempts to stabilize currency.

In 1533, Johann Widmanstetter, secretary to Pope Clement VII, explained Copernicus's heliocentric system to the Pope and two cardinals. The Pope was so pleased that he gave Widmanstetter a valuable gift. In 1535 Bernard Wapowski wrote a letter to a gentleman in Vienna, urging him to publish an enclosed almanac, which he claimed had been written by Copernicus. This is the only mention of a Copernicus almanac in the historical records. The "almanac" was likely Copernicus's tables of planetary positions. Wapowski's letter mentions Copernicus's theory about the motions of the earth. Nothing came of Wapowski's request, because he died a couple of weeks later.

Copernicus Monument in Warsaw designed by the Danish sculptor Bertel Thorvaldsen

Following the death of Prince-Bishop of Warmia Mauritius Ferber (1 July 1537), Copernicus participated in the election of his successor, Johannes Dantiscus (20 September 1537). Copernicus was one of four candidates for the post, written in at the initiative of Tiedemann Giese; but his candidacy was actually *pro forma*, since Dantiscus had earlier been named coadjutor bishop to Ferber. At first Copernicus maintained friendly relations with the new Prince-Bishop, assisting him medically in spring 1538 and accompanying him that summer on an inspection tour of Chapter holdings. But that autumn, their friendship was strained by suspicions over Copernicus's housekeeper, Anna Schilling, whom Dantiscus removed from Frombork in 1539.

In his younger days, Copernicus the physician had treated his uncle, brother and other chapter members. In later years he was called upon to attend the elderly bishops who in turn occupied the see of Warmia—Mauritius Ferber and Johannes Dantiscus – and, in 1539, his old friend

Tiedemann Giese, Bishop of Chełmno (Kulm). In treating such important patients, he sometimes sought consultations from other physicians, including the physician to Duke Albert and, by letter, the Polish Royal Physician.

Copernicus with medicinal plant

In the spring of 1541, Duke Albert summoned Copernicus to Königsberg to attend the Duke's counselor, George von

Kunheim, who had fallen seriously ill, and for whom the Prussian doctors seemed unable to do anything. Copernicus went willingly; he had met von Kunheim during negotiations over reform of the coinage. And Copernicus had come to feel that Albert himself was not such a bad person; the two had many intellectual interests in common. The Chapter readily gave Copernicus permission to go, as it wished to remain on good terms with the Duke, despite his Lutheran faith. In about a month the patient recovered, and Copernicus returned to Frombork. For a time, he continued to receive reports on von Kunheim's condition, and to send him medical advice by letter. Throughout this period of his life, Copernicus continued making astronomical observations and calculations, but only as his other responsibilities permitted and never in a professional capacity.

Some of Copernicus's close friends turned Protestant, but Copernicus never showed a tendency in that direction. The first attacks on him came from Protestants. Wilhelm Gnapheus, a Dutch refugee settled in Elbląg, wrote a comedy in Latin, Morosophus (The Foolish Sage), and staged it at the Latin school that he had established there. In the play, Copernicus was caricatured as a haughty, cold, aloof man who dabbled in astrology, considered himself inspired by God, and was rumored to have written a large work that was moldering in a chest.

Elsewhere Protestants were the first to react to news of Copernicus's theory. Melanchthon wrote:

Some people believe that it is excellent and correct to work out a thing as absurd as did that Sarmatian [i.e., Polish] astronomer who moves the earth and stops the sun. Indeed, wise rulers should have curbed such light-mindedness.

Nevertheless, in 1551, eight years after Copernicus's death, astronomer Erasmus Reinhold published, under the sponsorship of Copernicus's former military adversary, the Protestant Duke Albert, the Prussian Tables, a set of astronomical tables based on Copernicus's work. Astronomers and astrologers quickly adopted it in place of its predecessors.

- ROBERT STAWELL BALL

Heliocentrism

Some time before 1514 Copernicus made available to friends his "*Commentariolus*" ("Little Commentary"), a forty-page manuscript describing his ideas about the heliocentric hypothesis.[e] It contained seven basic assumptions (detailed below). Thereafter he continued gathering data for a more detailed work.

About 1532 Copernicus had basically completed his work on the manuscript of *Dē revolutionibus orbium coelestium*; but despite urging by his closest friends, he resisted openly publishing his views, not wishing—as he confessed—to risk the scorn "to which he would expose himself on account of the novelty and incomprehensibility of his theses."

In 1533, Johann Albrecht Widmannstetter delivered a series of lectures in Rome outlining Copernicus's theory. Pope

Clement VII and several Catholic cardinals heard the lectures and were interested in the theory. On 1 November 1536, Cardinal Nikolaus von Schönberg, Archbishop of Capua, wrote to Copernicus from Rome:

Some years ago word reached me concerning your proficiency, of which everybody constantly spoke. At that time I began to have a very high regard for you... For I had learned that you had not merely mastered the discoveries of the ancient astronomers uncommonly well but had also formulated a new cosmology. In it you maintain that the earth moves; that the sun occupies the lowest, and thus the central, place in the universe... Therefore with the utmost earnestness I entreat you, most learned sir, unless I inconvenience you, to communicate this discovery of yours to scholars, and at the earliest possible moment to send me your writings on the sphere of the universe together with the tables and whatever else you have that is relevant to this subject.

By then Copernicus's work was nearing its definitive form, and rumors about his theory had reached educated people all over Europe. Despite urgings from many quarters, Copernicus delayed publication of his book, perhaps from fear of criticism—a fear delicately expressed in the subsequent dedication of his masterpiece to Pope Paul III. Scholars disagree on whether Copernicus's concern was limited to possible astronomical and philosophical objections, or whether he was also concerned about religious objections.

The Book

Copernicus was still working on *De revolutionibus orbium coelestium* (even if not certain that he wanted to publish it) when in 1539 Georg Joachim Rheticus, a Wittenberg mathematician, arrived in Frombork. Philipp Melanchthon, a close theological ally of Martin Luther, had arranged for Rheticus to visit several astronomers and study with them. Rheticus became Copernicus's pupil, staying with him for two years and writing a book, Narratio prima (First Account), outlining the essence of Copernicus's theory. In 1542 Rheticus published a treatise on trigonometry by Copernicus (later included in the second book of *De revolutionibus*).

NICOLAI CO-
PERNICI TORINENSIS
DE REVOLVTIONIBVS ORBI-
um cœlestium, Libri VI.

.Habes in hoc opere iam recens nato, & ædito, studiose lector, Motus stellarum, tam fixarum, quàm erraticarum, cum ex ueteribus, tum etiam ex recentibus obseruationibus restitutos: & nouis insuper ac admirabilibus hypothesibus ornatos. Habes etiam Tabulas expeditissimas, ex quibus eosdem ad quoduis tempus quàm facillime calculare poteris. Igitur eme, lege, fruere.

Ἀγεωμέτρητος οὐδεὶς εἰσίτω.

Norimbergæ apud Ioh. Petreium,
Anno M. D. XLIII.

De revolutionibus, 1543, title page

Under strong pressure from Rheticus, and having seen the favorable first general reception of his work, Copernicus finally agreed to give *De revolutionibus* to his close friend, Tiedemann Giese, bishop of Chełmno (Kulm), to be delivered to Rheticus for printing by the German printer Johannes Petreius at Nuremberg (Nürnberg), Germany. While Rheticus initially supervised the printing, he had to leave Nuremberg before it was completed, and he handed over the task of supervising the rest of the printing to a Lutheran theologian, Andreas Osiander.

Osiander added an unauthorised and unsigned preface, defending the work against those who might be offended by the novel hypotheses. He explained that astronomers may find different causes for observed motions, and choose whatever is easier to grasp. As long as a hypothesis allows reliable computation, it does not have to match what a philosopher might seek as the truth.

Death

Toward the close of 1542, Copernicus was seized with apoplexy and paralysis, and he died at age 70 on 24 May 1543. Legend has it that he was presented with the final printed pages of his *Dē revolutionibus orbium coelestium* on the very day that he died, allowing him to take farewell of his life's work. He is reputed to have awoken from a stroke-induced coma, looked at his book, and then died peacefully.

Frombork Cathedral

Copernicus was reportedly buried in Frombork Cathedral, where archaeologists for over two centuries searched in vain for his remains. Efforts to locate the remains in 1802, 1909, 1939 and 2004 had come to nought. In August 2005, however, a team led by Jerzy Gąssowski, head of an archaeology and anthropology institute in Pułtusk, after scanning beneath the cathedral floor, discovered what they believed to be Copernicus's remains.

The find came after a year of searching, and the discovery was announced only after further research, on 3 November 2008. Gąssowski said he was "almost 100 percent sure it is Copernicus". Forensic expert Capt. Dariusz Zajdel of the Polish Police Central Forensic Laboratory used the skull to reconstruct a face that closely resembled the features—including a broken nose and a scar above the left eye—on a Copernicus self-portrait. The expert also determined that the skull belonged to a man who had died around age 70—

Copernicus's age at the time of his death.

Casket with Copernicus' remains, St. James' Cathedral Basilica, Olsztyn – March, 2010

The grave was in poor condition, and not all the remains of the skeleton were found; missing, among other things, was the lower jaw. The DNA from the bones found in the grave matched hair samples taken from a book owned by Copernicus which was kept at the library of the University of Uppsala in Sweden.

On 22 May 2010 Copernicus was given a second funeral in a Mass led by Józef Kowalczyk, the former papal nuncio to Poland and newly named Primate of Poland. Copernicus's remains were reburied in the same spot in Frombork Cathedral where part of his skull and other bones had been found. A black granite tombstone now identifies him as the founder of the heliocentric theory and also a church canon. The tombstone bears a representation of Copernicus's model of the solar system—a golden sun encircled by six of the planets.

Copernicus' 2010 grave, Frombork Cathedral

Copernican system

Predecessors

Philolaus (c. 480–385 BCE) described an astronomical system in which a Central Fire (different from the Sun) occupied the centre of the universe, and a counter-Earth, the Earth, Moon, the Sun itself, planets, and stars all revolved around it, in that order outward from the centre. Heraclides Ponticus (387–312 BCE) proposed that the Earth rotates on its axis. Aristarchus of Samos (310 BCE – c. 230 BCE) was the first to advance a theory that the earth orbited the sun. Further mathematical details of Aristarchus' heliocentric system were worked out around 150 BC by the Hellenistic astronomer Seleucus of Seleucia. Though Aristarchus' original text has been lost, a reference in Archimedes' book The Sand Reckoner (Archimedis Syracusani Arenarius & Dimensio Circuli) describes a work by Aristarchus in which he advanced the heliocentric model. Archimedes wrote:

You (King Gelon) are aware the 'universe' is the name given by most astronomers to the sphere the center of which is the center of the Earth, while its radius is equal to the straight line between the center of the Sun and the center of the Earth. This is the common account as you have heard from astronomers. But Aristarchus has brought out a book consisting of certain hypotheses, wherein it appears, as a consequence of the assumptions made, that the universe is many times greater than the 'universe' just mentioned. His hypotheses are that the fixed stars and the Sun remain unmoved, that the Earth revolves about the Sun on the circumference of a circle, the Sun lying in the middle of the Floor, and that the sphere of the fixed stars, situated about the same center as the Sun, is so great that the circle in which he supposes the Earth to revolve bears such a proportion to the distance of the fixed stars as the center of the sphere bears to its surface.

— The Sand Reckoner

1735 epitaph, Frombork Cathedral.
A 1580 epitaph had been destroyed.

Copernicus cited Aristarchus of Samos in an early (unpublished) manuscript of *De Revolutionibus*

(which still survives), though he removed the reference from his final published manuscript.

Some technical details of Copernicus's system closely resembled those developed earlier by the Islamic astronomers Naṣīr al-Dīn al-Ṭūsī and Ibn al-Shāṭir, both of whom retained a geocentric model. Aryabhata (476–550), in his magnum opus Aryabhatiya (499), propounded a planetary model in which the Earth was taken to be spinning on its axis and the periods of the planets were given with respect to the Sun. He accurately calculated many astronomical constants, such as the periods of the planets, times of the solar and lunar eclipses, and the instantaneous motion of the Moon.

At the Maragha observatory, Najm al-Dīn al-Qazwīnī al-Kātibī (died 1277), in his

Hikmat al-'Ain, wrote an argument for a heliocentric model, but later abandoned the model. Qutb al-Din Shirazi (born 1236) also discussed the possibility of heliocentrism, but rejected it. Ibn al-Shatir (born 1304) developed a geocentric system that employed mathematical techniques, such as the ***Tusi couple*** and Urdi lemma, that were almost identical to those Nicolaus Copernicus later employed in his heliocentric system, implying that its mathematical model was influenced by the Maragha school.

Nilakantha Somayaji (1444–1544), in his Aryabhatiyabhasya, a commentary on Aryabhata's Aryabhatiya, developed a computational system for a partially heliocentric planetary model, in which the planets orbit the Sun, which in turn orbits the Earth, similar to the Tychonic system later proposed by Tycho Brahe in the late 16th century. In the Tantrasangraha (1500), he further revised his planetary system, which was mathematically more accurate at predicting the heliocentric orbits of the interior planets than both the Tychonic and Copernican models.

The prevailing theory in Europe during Copernicus's lifetime was the one that Ptolemy published in his *Almagest* circa 150 CE; the Earth was the stationary center of the universe. Stars were embedded in a large outer sphere which rotated rapidly, approximately daily, while each of the planets, the Sun, and the Moon were embedded in their own, smaller spheres. Ptolemy's system employed devices, including epicycles, deferents and equants, to account for observations that the paths of these bodies differed from simple, circular orbits centered on the Earth.

Copernicus's vision of the universe in *Dē revolutionibus orbium coelestium*

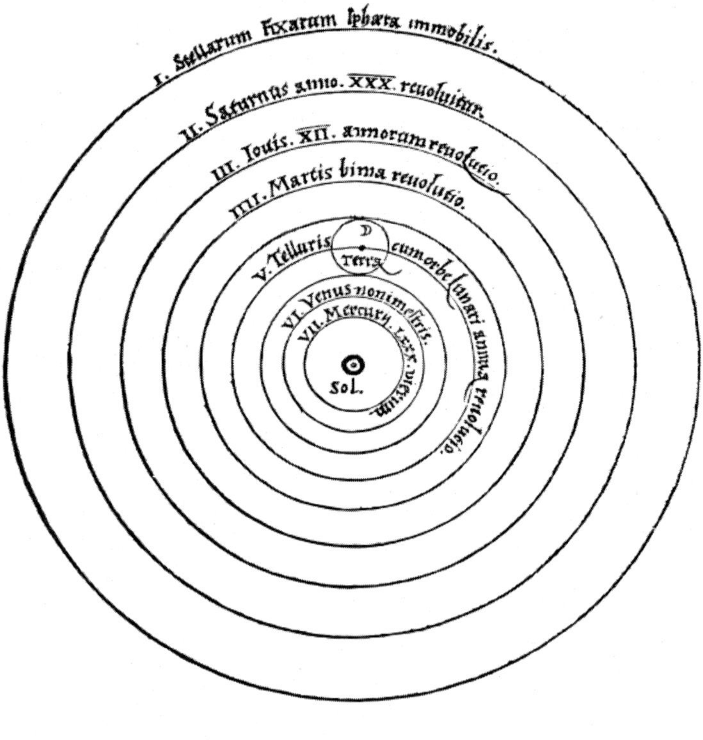

Copernicus's vision of the universe in
Dē revolutionibus orbium coelestium

Copernicus's major theory was published in *Dē revolutionibus orbium coelestium* (On the Revolutions of the Celestial Spheres), in the year of his death, 1543, though he had formulated the theory several decades earlier.

Copernicus's "*Commentariolus*" summarized his heliocentric theory. It listed the "assumptions" upon which the theory was based, as follows:

1. There is no one center of all the celestial circles or spheres.

2. The center of the earth is not the center of the universe, but only of gravity and of the lunar sphere.

3. All the spheres revolve about the sun as their midpoint, and therefore the sun is the center of the universe.

4. The ratio of the earth's distance from the sun to the height of the firmament

(outermost celestial sphere containing the stars) is so much smaller than the ratio of the earth's radius to its distance from the sun that the distance from the earth to the sun is imperceptible in comparison with the height of the firmament.

5. Whatever motion appears in the firmament arises not from any motion of the firmament, but from the earth's motion. The earth together with its circumjacent elements performs a complete rotation on its fixed poles in a daily motion, while the firmament and highest heaven abide unchanged.

6. What appear to us as motions of the sun arise not from its motion but from the motion of the earth and our sphere, with which we revolve about the sun like any other planet. The earth has, then, more than one motion.

7. The apparent retrograde and direct motion of the planets arises not from their motion but from the earth's. The motion of

the earth alone, therefore, suffices to explain so many apparent inequalities in the heavens.

De revolutionibus itself was divided into six parts, called "books":

- General vision of the heliocentric theory, and a summarized exposition of his idea of the World
- Mainly theoretical, presents the principles of spherical astronomy and a list of stars (as a basis for the arguments developed in the subsequent books)
- Mainly dedicated to the apparent motions of the Sun and to related phenomena
- Description of the Moon and its orbital motions
- Exposition of the motions in longitude of the non-terrestrial planets

- Exposition of the motions in latitude of the non-terrestrial planets

Successors

Georg Joachim Rheticus could have been Copernicus's successor, but did not rise to the occasion. Erasmus Reinhold could have been his successor, but died prematurely. The first of the great successors was Tycho Brahe (though he did not think the earth orbited the sun), followed by Johannes Kepler, who had worked as Tycho's assistant in Prague.

Despite the near universal acceptance later of the heliocentric idea (though not the epicycles or the circular orbits), Copernicus's theory was originally slow to catch on. Scholars hold that sixty years after the publication of The Revolutions there were only around 15 astronomers espousing Copernicanism in all of Europe: "Thomas Digges and Thomas Harriot in England; Giordano Bruno and Galileo Galilei in Italy; Diego Zuniga in Spain;

Simon Stevin in the Low Countries; and in Germany, the largest group – Georg Joachim Rheticus, Michael Maestlin, Christoph Rothmann (who may have later recanted), and Johannes Kepler." Additional possibilities are Englishman William Gilbert, along with Achilles Gasser, Georg Vogelin, Valentin Otto, and Tiedemann Giese.

Arthur Koestler, in his popular book The Sleepwalkers, asserted that Copernicus's book had not been widely read on its first publication. This claim was trenchantly criticised by Edward Rosen, and has been decisively disproved by Owen Gingerich, who examined nearly every surviving copy of the first two editions and found copious marginal notes by their owners throughout many of them. Gingerich published his conclusions in 2004 in The Book Nobody Read.

The intellectual climate of the time "remained dominated by Aristotelian philosophy and the corresponding Ptolemaic astronomy. At that time there was no reason to accept the Copernican theory, except for its mathematical simplicity [by avoiding using the equant in determining planetary positions]." Tycho Brahe's system ("that the earth is stationary, the sun revolves about the earth, and the other planets revolve about the sun") also directly competed with Copernicus's. It was only a half century later with the work of Kepler and Galileo that any substantial evidence defending Copernicanism appeared, starting "from the time when Galileo formulated the principle of inertia...[which] helped to explain why everything would not fall off the earth if it were in motion." It was not until "after Isaac Newton formulated the universal law of gravitation and the laws of mechanics [in his 1687 Principia], which unified terrestrial and celestial mechanics, was the heliocentric view generally accepted."

Controversy

Only mild controversy was the immediate result of the 1543 publication of Copernicus's book. At the Council of Trent (1545–63) neither Copernicus's theory nor calendar reform (which would later use tables deduced from Copernicus's calculations) were discussed. It has been much debated why it was not until six decades after the publication of *De revolutionibus* that the Catholic Church took any official action against it, even the efforts of Tolosani going unheeded. Catholic side opposition only commenced seventy-three years later, when it was occasioned by Galileo.

Tolosani

The first notable to move against Copernicanism was the Magister of the Holy Palace (i.e., the Catholic Church's chief censor), Dominican Bartolomeo

Spina, who "expressed a desire to stamp out the Copernican doctrine". But with Spina's death in 1546, his cause fell to his friend, the well-known theologian-astronomer, the Dominican Giovanni Maria Tolosani of the Convent of St. Mark in Florence. Tolosani had written a treatise on reforming the calendar (in which astronomy would play a large role) and had attended the Fifth Lateran Council (1512–17) to discuss the matter. He had obtained a copy of *De Revolutionibus* in 1544. His denunciation of Copernicanism was written a year later, in 1545, in an appendix to his unpublished work, On the Truth of Sacred Scripture.

Emulating the rationalistic style of Thomas Aquinas, Tolosani sought to refute Copernicanism by philosophical argument. Copernicanism was absurd, according to Tolosani, because it was scientifically unproven and unfounded. First, Copernicus had assumed the motion of the Earth but offered no physical theory whereby one would deduce this motion. (No one realized that the investigation into Copernicanism would result in a rethinking of the entire field of physics.) Second, Tolosani charged that Copernicus's thought process was backwards. He held that Copernicus had come up with his idea and then sought phenomena that would support it, rather than observing phenomena and deducing from them the idea of what caused them. In this, Tolosani was linking Copernicus's mathematical equations with the practices of the Pythagoreans (whom Aristotle had made arguments against, which were later picked up by Thomas Aquinas). It was argued that mathematical numbers were a mere product of the intellect without any physical reality, and as such could not provide physical causes in the investigation of nature.

Some astronomical hypotheses at the time (such as epicycles and eccentrics) were seen as mere mathematical devices to adjust calculations of where the heavenly bodies would appear, rather than an explanation of the cause of those motions. (As Copernicus still maintained the idea of perfectly spherical orbits, he relied on epicycles.) This "saving the phenomena" was seen as proof that astronomy and mathematics could not be taken as serious means to determine physical causes. Tolosani invoked this view in his final critique of Copernicus, saying that his biggest error was that he had started with "inferior" fields of science to make pronouncements about "superior" fields. Copernicus had used mathematics and astronomy to postulate about physics and cosmology, rather than beginning with the accepted principles of physics and cosmology to determine things about astronomy and mathematics. Thus Copernicus seemed to be undermining the whole system of the philosophy of science at the time. Tolosani held that Copernicus had fallen into philosophical error because he had not been versed in physics and logic; anyone without such knowledge would make a poor astronomer and be unable to distinguish truth from falsehood. Because Copernicanism had not met the criteria for scientific truth set out by Thomas Aquinas, Tolosani held that it could only be viewed as a wild unproven theory.

Tolosani recognized that the Ad Lectorem preface to Copernicus's book was not actually by him. Its thesis that astronomy as a whole would never be able to make truth claims was rejected by Tolosani (though he still held that Copernicus's attempt to describe physical reality had been faulty); he found it ridiculous that Ad Lectorem had been included in the book (unaware that Copernicus had not authorized its inclusion). Tolosani wrote: "By means of these words [of the Ad Lectorem], the foolishness of this book's author is rebuked. For by a foolish effort he [Copernicus] tried to revive the weak Pythagorean opinion [that the element of fire was at the center of the Universe], long ago deservedly destroyed, since it is expressly contrary to human reason and also opposes holy writ. From this situation, there could easily arise disagreements between Catholic expositors of holy scripture and those who might wish to adhere obstinately to this false opinion." Tolosani declared: "Nicolaus Copernicus neither read nor understood the arguments of Aristotle the philosopher and Ptolemy the astronomer." Tolosani wrote that Copernicus "is expert indeed in the sciences of mathematics and astronomy, but he is very deficient in the sciences of physics and logic. Moreover, it appears that he is unskilled with regard to [the interpretation of] holy scripture, since he contradicts several of its principles, not without danger of infidelity to himself and the readers of his book. ...his arguments have no force and can very easily be taken apart. For it is stupid to contradict an opinion accepted by everyone over a very long time for the strongest reasons, unless the impugner uses more powerful and insoluble demonstrations and completely dissolves the opposed reasons. But he does not do this in the least."

Tolosani declared that he had written against Copernicus "for the purpose of preserving the truth to the common advantage of the Holy Church." Despite this, his work remained unpublished and there is no evidence that it received serious consideration. Robert Westman describes it as becoming a "dormant" viewpoint with "no audience in the Catholic world" of the late sixteenth century, but also notes that there is some evidence that it did become known to Tommaso Caccini, who would criticize Galileo in a sermon in December 1613.

Theology

Tolosani may have criticized the Copernican theory as scientifically unproven and unfounded, but the theory also conflicted with the theology of the time, as can be seen in a sample of the works of John Calvin. In his Commentary on Genesis he said that "We indeed are not ignorant that the circuit of the heavens is finite, and that the earth, like a little globe, is placed in the centre." In his commentary on Psalms 93:1 he states that "The heavens revolve daily, and, immense as is their fabric and inconceivable the rapidity of their revolutions, we experience no concussion.... How could the earth hang suspended in the air were it not upheld by God's hand? By what means could it maintain itself unmoved, while the heavens above are in constant rapid motion, did not its Divine Maker fix and establish it." One sharp point of conflict between Copernicus's theory and the Bible concerned the story of the Battle of Gibeon in the Book of Joshua where the Hebrew forces were winning but whose opponents were likely to escape once night fell. This is averted by Joshua's prayers causing the sun and the moon to stand still. Martin Luther once made a remark about Copernicus, although without mentioning his name. According to Anthony Lauterbach, while eating with Martin Luther the topic of Copernicus arose during dinner on 4 June 1539 (in the same year as professor George Joachim Rheticus of the local University had been granted leave to visit him). Luther is said to have remarked "So it goes now. Whoever wants to be clever must agree with nothing others esteem. He must do something of his own. This is what that fellow does who wishes to turn the whole of astronomy upside down. Even in these things that are thrown into disorder I believe the Holy Scriptures, for Joshua commanded the sun to stand still and not the earth." These remarks were made four years before the publication of On the Revolutions of the Heavenly Spheres and a year before Rheticus' Narratio Prima. In John Aurifaber's account of the conversation Luther calls Copernicus "that fool" rather than "that fellow", this version is viewed by historians as less reliably sourced.

Luther's collaborator Philipp Melanchthon also took issue with Copernicanism. After receiving the first pages of Narratio Prima from Rheticus himself, Melanchthon wrote to Mithobius (physician and mathematician Burkard Mithob of Feldkirch) on 16 October 1541 condemning the theory and calling for it to be repressed by governmental force, writing "certain people believe it is a marvelous achievement to extol so crazy a thing, like that Polish astronomer who makes the earth move and the sun stand still. Really, wise governments ought to repress impudence of mind." It had appeared to Rheticus that Melanchton would understand the theory and would be open to it. This was because Melanchton had taught Ptolemaic astronomy and had even recommended his friend Rheticus to an appointment to the Deanship of the Faculty of Arts & Sciences at the

University of Wittenberg after he had returned from studying with Copernicus.

Rheticus' hopes were dashed when six years after the publication of *De Revolutionibus* Melanchthon published his Initia Doctrinae Physicae presenting three grounds to reject Copernicanism. These were "the evidence of the senses, the thousand-year consensus of men of science, and the authority of the Bible". Blasting the new theory Melanchthon wrote "Out of love for novelty or in order to make a show of their cleverness, some people have argued that the earth moves. They maintain that neither the eighth sphere nor the sun moves, whereas they attribute motion to the other celestial spheres, and also place the earth among the heavenly bodies. Nor were these jokes invented recently. There is still extant Archimedes' book on The sand-reckoner; in which he reports that Aristarchus of Samos propounded the paradox that the sun stands still and the earth revolves around the sun. Even though subtle experts institute many investigations for the sake of exercising their ingenuity, nevertheless public proclamation of absurd opinions is indecent and sets a harmful example." Melanchthon went on to cite Bible passages and then declare "Encouraged by this divine evidence, let us cherish the truth and let us not permit ourselves to be alienated from it by the tricks of those who deem it an intellectual honor to introduce confusion into the arts." In the first edition of Initia Doctrinae Physicae, Melanchthon even questioned Copernicus's character claiming his motivation was "either from love of novelty or from desire to appear clever", these more personal attacks were largely removed by the second edition in 1550.

Another Protestant theologian who took issue with Copernicus was John Owen who declared that "the late hypothesis, fixing the sun as in the centre of the world' was 'built on fallible phenomena, and advanced by many arbitrary presumptions against evident testimonies of Scripture."

In Roman Catholic circles, German Jesuit Nicolaus Serarius was one of the first to write against Copernicus's theory as heretical, citing the Joshua passage, in a work published in 1609–1610, and again in a book in 1612. In his 12 April 1615 letter to a Catholic defender of Copernicus, Paolo Antonio Foscarini, Catholic Cardinal Robert Bellarmine condemned Copernican theory, writing "...not only the Holy Fathers, but also the modern commentaries on Genesis, the Psalms, Ecclesiastes, and Joshua, you will find all agreeing in the literal interpretation that the sun is in heaven and turns around the earth with great speed, and that the earth is very far from heaven and sits motionless at the center of the world...Nor can one answer that this is not a matter of faith, since if it is not a matter of faith 'as regards the topic,' it is a matter of faith 'as regards the speaker': and so it would be heretical to say that Abraham did not have two children and Jacob twelve, as well as to say that Christ was not born of a virgin, because both are said by the Holy Spirit through the mouth of prophets and apostles."

Ingoli

Perhaps the most influential opponent of the Copernican theory was Francesco Ingoli, a Catholic priest. Ingoli wrote a January 1616 essay to Galileo presenting more than twenty arguments against the Copernican theory. Though "it is not certain, it is probable that he [Ingoli] was commissioned by the Inquisition to write an expert opinion on the controversy", (after the Congregation of the Index's decree against Copernicanism on 5 March 1616, Ingoli was officially appointed its consultant). Galileo himself was of the opinion that the essay played an important role in the rejection of the theory by church authorities, writing in a later letter to Ingoli that he was concerned that people thought the theory was rejected because Ingoli was right. Ingoli presented five physical arguments against the theory, thirteen mathematical arguments (plus a separate discussion of the sizes of stars), and four theological arguments. The physical and mathematical arguments were of uneven quality, but many of them came directly from the writings of Tycho Brahe, and Ingoli repeatedly cited Brahe, the leading astronomer of the era. These included arguments about the effect of a moving earth on the trajectory of projectiles, and about parallax and Brahe's argument that the Copernican theory required that stars be absurdly large. Two of Ingoli's theological issues with the Copernican theory were "common Catholic beliefs not directly traceable to Scripture: the doctrine that hell is located at the center of Earth and is most distant from heaven; and the explicit assertion that Earth is motionless in a hymn sung on

Tuesdays as part of the Liturgy of the Hours of the Divine Office prayers regularly recited by priests." Ingoli cited Robert Bellarmine in regards to both of these arguments, and may have been trying to convey to Galileo a sense of Bellarmine's opinion. Ingoli also cited Genesis 1:14 where God places "lights in the firmament of the heavens to divide the day from the night." Ingoli did not think the central location of the sun in the Copernican theory was compatible with it being described as one of the lights placed in the firmament. Like previous commentators Ingoli also pointed to the passages about the Battle of Gibeon. He dismissed arguments that they should be taken metaphorically, saying "Replies which assert that Scripture speaks according to our mode of understanding are not satisfactory: both because in explaining the Sacred Writings the rule is always to preserve the literal sense, when it is possible, as it is in this case; and also because all the [Church] Fathers unanimously take this passage to mean that the sun which was truly moving stopped at Joshua's request. An interpretation which is contrary to the unanimous consent of the Fathers is condemned by the Council of Trent, Session IV, in the decree on the edition and use of the Sacred Books. Furthermore, although the Council speaks about matters of faith and morals, nevertheless it cannot be denied that the Holy Fathers would be displeased with an interpretation of Sacred Scriptures which is contrary to their common agreement." However, Ingoli closed the essay by suggesting Galileo respond primarily to the better of his physical and mathematical arguments rather than to his theological arguments, writing "Let it be your choice to respond to this either entirely of in part—clearly at least to the mathematical and physical arguments, and not to all even of these, but to the more weighty ones." When Galileo wrote a letter in reply to Ingoli years later, he in fact only addressed the mathematical and physical arguments.

In March 1616, in connection with the Galileo affair, the Roman Catholic Church's Congregation of the Index issued a decree suspending *De revolutionibus* until it could be "corrected," on the grounds of ensuring that Copernicanism, which it described as a "false Pythagorean doctrine, altogether contrary to the Holy Scripture," would not "creep any further to the prejudice of Catholic truth." The corrections consisted largely of removing or altering wording that the spoke of heliocentrism as a fact, rather than a hypothesis. The corrections were made based largely on work by Ingoli.

Galileo

On the orders of Pope Paul V, Cardinal Robert Bellarmine gave Galileo prior notice that the decree was about to be issued, and warned him that he could not "hold or defend" the Copernican doctrine. The corrections to *De revolutionibus*, which omitted or altered nine sentences, were issued four years later, in 1620.

In 1633 Galileo Galilei was convicted of grave suspicion of heresy for "following the position of Copernicus, which is contrary to the true sense and authority of Holy Scripture", and was placed under house arrest for the rest of his life.

At the instance of Roger Boscovich, the Catholic Church's 1758 Index of Prohibited Books omitted the general prohibition of works defending heliocentrism, but retained the specific prohibitions of the original uncensored versions of *De revolutionibus* and Galileo's Dialogue Concerning the Two Chief World Systems. Those prohibitions were finally dropped from the 1835 Index.

- ROBERT STAWELL BALL

Nationality

There has been discussion of Copernicus's nationality and of whether, in fact, it is meaningful to ascribe to him a nationality in the modern sense. But there are some facts that should be considered when talking about Copernicus's nationality.

Nicolaus Copernicus was born and raised in Royal Prussia, a semiautonomous and polyglot region under rule by the Kingdom of Poland. Copernicus was the child of German-speaking parents and grew up with German as his mother tongue. His alma mater was the University of Kraków in Poland. When he later studied in Italy, Copernicus joined the "German Nation", the student organization for German speakers at the University of Bologna. His family stood against the Teutonic Order and actively supported the city of Toruń during the Thirteen Years' War. The father of Nicolaus Copernicus lent money to King Casimir IV Jagiellon of Poland to finance the war against the Teutonic Knights, but the inhabitants of Royal Prussia also resisted the Polish crown's efforts for greater control over the region.

Replica of Warsaw's Copernicus
Monument, in Montreal, Canada

Bust by Schadow, 1807,
in the Walhalla memorial

Encyclopædia Britannica, Encyclopedia Americana, The Concise Columbia Encyclopedia, The Oxford World Encyclopedia, and World Book Encyclopedia refer to Copernicus as a "Polish astronomer". Sheila Rabin, writing in the Stanford Encyclopedia of Philosophy, describes Copernicus as a "child of a German family [who] was a subject of the Polish crown", while Martin Weissenbacher writes that Copernicus's father was a Germanized Pole.

Historian Michael Burleigh describes the nationality debate as a "totally insignificant battle" between German and Polish scholars during the interwar period. Polish astronomer Konrad Rudnicki calls the discussion a "fierce scholarly quarrel in... times of nationalism" and describes Copernicus as an inhabitant of a German-speaking territory that belonged to Poland, himself being of mixed Polish-German extraction.

Poet Czesław Miłosz describes the debate as an "absurd" projection of a modern understanding of nationality onto Renaissance people, who identified with their home territories rather than with a nation.

Similarly, historian Norman Davies writes that Copernicus, as was common in his era, was "largely indifferent" to nationality, being a local patriot who considered himself "Prussian".

Miłosz and Davies both write that Copernicus had a German-language cultural background, while his working language was Latin in accord with the usage of the time. Additionally, according to Davies, "there is ample evidence that he knew the Polish language". Davies concludes: "Taking everything into consideration, there is good reason to regard him both as a German and as a Pole: and yet, in the sense that modern nationalists understand it, he was neither."

Commemoration

Copernicium

On 14 July 2009, the discoverers, from the Gesellschaft für Schwerionenforschung in Darmstadt, Germany, of chemical element 112 (temporarily named ununbium) proposed to the International Union of Pure and Applied Chemistry that its permanent name be "copernicium" (symbol Cn). "After we had named elements after our city and our state, we wanted to make a statement with a name that was known to everyone," said Hofmann. "We didn't want to select someone who was a German. We were looking world-wide." On the 537th anniversary of his birthday the official naming was released to the public.

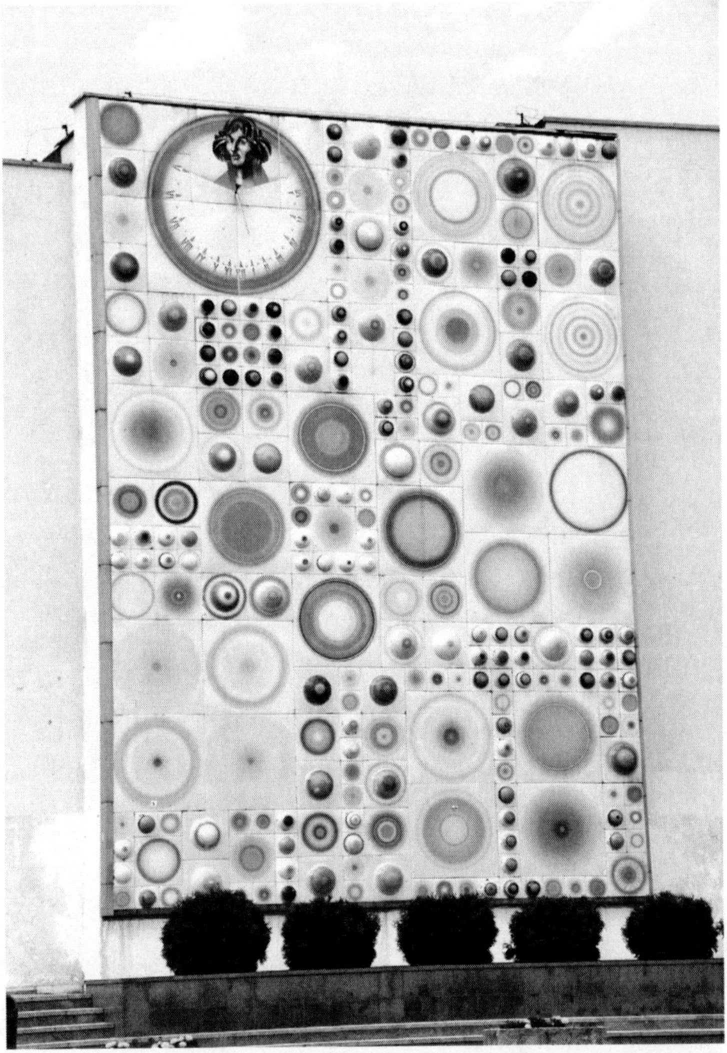

Enamel mosaic by Stefan Knapp, 1972, on building of Nicolaus Copernicus University, in Toruń, Poland.

- ROBERT STAWELL BALL

55 Cancri A

In July 2014 the International Astronomical Union launched a process for giving proper names to certain exoplanets and their host stars. The process involved public nomination and voting for the new names. In December 2015, the IAU announced the winning name for 55 Cancri A was Copernicus.

Veneration

Copernicus is honored, together with Johannes Kepler, in the liturgical calendar of the Episcopal Church (USA), with a feast day on 23 May.

Influence

Contemporary literary and artistic works inspired by Copernicus

- *Mover of the Earth, Stopper of the Sun* for symphony orchestra (overture), written by composer Svitlana Azarova commissioned by ONDIF
- Doctor Copernicus, a 1975 novel by John Banville, sketches the life of Copernicus and the 16th-century world in which he lived.

U.S. postage stamp on 500th anniversary
of Copernicus's birth (1973)

ABOUT THE AUTHOR

ROBERT STAWELL BALL
(July 1, 1840 – November 25, 1913)

Sir Robert Stawell Ball was an Irish astronomer who founded the screw theory.

He was the son of naturalist Robert Ball and Amelia Gresley Hellicar. He was born in Dublin.

Ball worked for Lord Rosse from 1865 to 1867. In 1867 he became Professor of Applied Mathematics at the Royal College of Science in Dublin. There he lectured on mechanics and published an elementary account of the science.

In 1874 Ball was appointed Royal Astronomer of Ireland and Andrews Professor of Astronomy in the University of Dublin at Dunsink Observatory.

Ball contributed to the science of kinematics by delineating the screw displacement:

When Ball and the screw theorists speak of screws they no longer mean actual cylindrical objects with helical threads cut into them but the possible motion of any body whatsoever, including that of the screw independently of the nut.

Ball's treatise *The Theory of Screws* (1876) is now in the public domain. His work on screw dynamics earned him in 1879 the

Cunningham Medal of the Royal Irish Academy.

In 1882 Popular Science Monthly carried his article "*A Glimpse through the Corridors of Time*". The following year it carried his two-part article on "*The Boundaries of Astronomy*".

Ball expounded the tides in *Time and Tide*: a Romance of the Moon In 1892 he was appointed Lowndean Professor of Astronomy and Geometry at Cambridge University at the same time becoming director of the Cambridge Observatory. He was a fellow of King's College, Cambridge.

In 1900 Cambridge University Press published *A Treatise on the Theory of Screws*. That year he also published *The Story of the Heavens*. Much in the limelight, he stood as President of the Quaternion Society. He was also President of the Mathematical Association in 1900.

In 1908 he published *A Treatise on Spherical Astronomy*, which is a textbook on astronomy starting from spherical trigonometry and the celestial sphere, considering atmospheric refraction and aberration of light, and introducing basic use of a generalised instrument.

His work *The Story of the Heavens* is mentioned in the "Ithaka" chapter of Ulysses. His lectures, articles and books (e.g. *Starland* and *The Story of the Heavens*) were mostly popular and simple in style.

He died in Cambridge and was buried at the Parish of the Ascension Burial Ground in Cambridge, with his wife Lady Francis Elizabeth Ball. Their children were: Frances Amelia, Robert Steele, William Valentine (later Sir), Mary Agnetta, Charles Rowan Hamilton, and Randall Gresley (later Colonel).

LECTURES

Ball became celebrated for his popular lectures on science. He gave an estimated 2500 lectures between 1875 and 1910 in towns and cities across Britain and Ireland. In 1892, 1898 and 1900 he was invited to deliver the Royal Institution Christmas Lecture. *Astronomy; Astronomy* and *Great Chapters from the Book of Nature.*

FEEDBACK

Now that you have read the book ...

Was it interesting?

Did you enjoy what you wanted to read?
Was there any room for improvement?

Let us know at:

http://www.diamondbooks.ca/feedback

Your feedback is highly appreciated.

Thank you!

NOTES

NOTES

Made in United States
Cleveland, OH
06 November 2026

47647251R00087